U.S. Fire Administration

Fire-Related Firefighter Injuries in 2004

February 2008

 FEMA

U.S. Fire Administration

Mission Statement

As an entity of the Federal Emergency Management Agency (FEMA), the mission of the U.S. Fire Administration (USFA) is to reduce life and economic losses due to fire and related emergencies, through leadership, advocacy, coordination, and support. We serve the Nation independently, in coordination with other Federal agencies, and in partnership with fire protection and emergency service communities. With a commitment to excellence, we provide public education, training, technology, and data initiatives.

Contents

FIGURES

TABLES

FIRE-RELATED FIREFIGHTER INJURIES IN 2004

Introduction

Every occupation brings degrees of safety risk, and one of the higher risk jobs is firefighting. At the fire scene or on the way to a fire, vehicle crash, or explosion, or even while training, firefighters face a relatively high chance of being injured and possibly killed. Each year, tens of thousands of firefighters are injured while fighting fires, rescuing people, responding to emergency medical incidents, responding to hazardous materials incidents, or training for their job. While the majority of injuries are minor, a significant number are debilitating and career-ending. Such injuries exact a great toll on the fabric of the fire service.

From the need to adjust staffing levels and rotations to accommodate injuries, to the focus of the fire service on injury prevention, injuries and their prevention are a primary concern. In addition, the fire service has done much to improve firefighter safety. Firefighter health and safety initiatives, incident command structure, training, and protective gear are but a few areas where time, energy, and resources have been well-spent. Nonetheless, firefighting is, by its very nature, a hazardous profession. Injuries can and do occur.

This report presents the details of firefighter injuries sustained at or responding to a fire incident, focusing on 2004 data. These injuries may be the result of operations at the fire scene or responding to or returning from an incident. Confirmation of (or opposition to) previous reported trends of firefighter injuries are noted where appropriate. Most of the statistics presented are from analyses of the 2004 National Fire Incident Reporting System (NFIRS) Version 5.0 data.

Methodology

Fire-Related Firefighter Injuries in 2004 relies on data from the Nation's largest fire incident database, NFIRS, and on independent research from a variety of public and private organizations including the National Fire Protection Association (NFPA).

Data Source

The fire-related findings in this report are based primarily on analysis of NFIRS fire incident data for 2004. NFIRS is a voluntary data collection system administered by the United States Fire Administration (USFA), an agency under the Department of Homeland Security (DHS). The participating fire departments include career, volunteer, and combination departments that serve communities ranging from rural hamlets to the largest cities. Participation in NFIRS is State-based and voluntary. Not all States participate and, for those that do, reported fire incidents do not reflect all of a State's fire activity. Also, not all recorded information is complete. For example, not all data items are filled out and there are known instances of departments reporting fire information but not information on the associated casualties. In the former case, the data are examined and analyses are pursued only if sufficient information is available. In the latter case, departments with known or suspected data inadequacies are removed from the dataset.[1] Nevertheless, each year of NFIRS fire incident data represents the participation of one-third to one-half of the Nation's fire departments and contains approximately 800,000 records, each representing a separate fire incident.

NFIRS underwent a substantial transition in 1999 to Version 5.0. The system has continued to accept legacy data (Version 4.1); by 2004, less than 11 percent of the data was legacy data. This report relies only on data from the current Version 5.0.

Percentages in Fire Data

Since the data set is incomplete and represents only a sample of American fire departments, many of the numbers in this analysis are percentages rather than raw totals or absolute numbers. In making national estimates of the fire problem, unknown or undetermined data in the NFIRS database are not ignored. Unknown data occur when the information in nonrequired data collection items in NFIRS is not provided (left blank), the coding provided is invalid, or the information is noted as "undetermined." The general approach taken in this report is to include unknown data as a separate category (i.e., "unspecified, unknown") as a measure of the validity of the data element.

Rounding

Percentages on each figure and table are rounded to one decimal point. Textual discussions cite these percentages as whole numbers. Thus, both 12.5 percent and 13.4 percent are rounded to 13 percent; 13.5 percent is rounded to 14 percent.

Counting Injuries and Injury-Producing Incidents

COUNTING INJURIES. In NFIRS, firefighter fire injuries are derived from the firefighter casualty records with an associated incident type in the 100 series (fire incidents). Firefighter injuries that result from any Version 5.0 fire incident, including mutual-aid incidents, are included in the injury counts. As only Version 5.0 records are considered, firefighter injuries associated with incident type 110 (a conversion code for the version 4.1 incidents) are not included. Injuries are determined by the severity of the casualty. Severity of firefighter injuries are coded on a 7-level scale ranging from a hazardous exposure to death. Severity 1 (report only, including exposures) and severity 7 (death) are not included in injury counts; all other severity values are counted as injuries, including any injury records where the severity code may not be captured. For a substantial number of incidents, there is insufficient information available to match aid-given records to aid-received records. As a result, analyses of injuries per fire, exclude both injuries from mutual-aid incidents and the mutual-aid incidents themselves.

COUNTING INJURY-PRODUCING INCIDENTS. Firefighter injury incidents are those incidents that have an associated firefighter casualty record that meets the injury criteria above with one major exception to prevent the double counting of aid-given incidents: In the case of mutual or automatic aid given or received, the incident record associated with aid given is excluded in the injury incident counts.

---

[1] In particular, data from New York City has been excluded from this analysis of firefighter fire-related injuries, as no firefighter injuries associated with fires were reported for 2004.

Injury-Producing Incidents

Twice as many firefighters are injured each year performing fireground duties as there are fire injuries to the civilian population (36,880 versus 17,875 in 2004).[2, 3] In all, 75,840 firefighters were injured in 2004 while on duty.[4] The 10-year trends, however, in both total firefighter injuries and fireground injuries continued downward—16 and 23 percent, respectively—as shown in Figure 1. Because the number of reported fires has also fallen over the 10-year period, on a per-fire basis the downward trend was a much more modest 3 percent (Figure 2).

Figure 1. Trends in Firefighter Injuries, 1995–2004

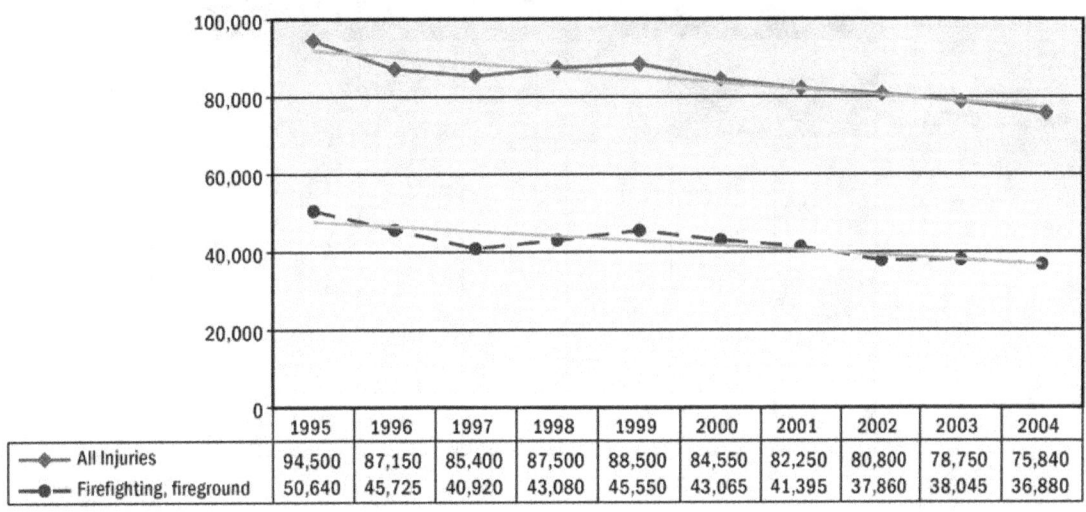

	1995	1996	1997	1998	1999	2000	2001	2002	2003	2004
All Injuries	94,500	87,150	85,400	87,500	88,500	84,550	82,250	80,800	78,750	75,840
Firefighting, fireground	50,640	45,725	40,920	43,080	45,550	43,065	41,395	37,860	38,045	36,880

Source: NFPA, annual National Fire Experience Surveys.

Figure 2. Trends in Firefighter Fireground Injuries per Thousand Fires, 1995–2004

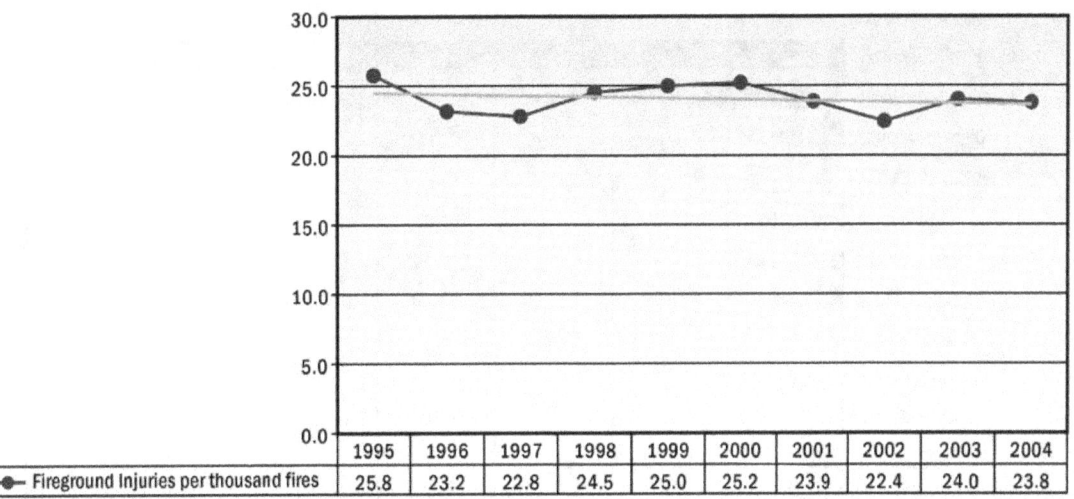

	1995	1996	1997	1998	1999	2000	2001	2002	2003	2004
Fireground Injuries per thousand fires	25.8	23.2	22.8	24.5	25.0	25.2	23.9	22.4	24.0	23.8

Source: NFPA, annual National Fire Experience Surveys.

---

[2] Injury estimates are from the NFPA's annual surveys. The 2004 estimate for firefighter fireground injuries is 36,880, to which a portion of the injuries categorized as responding to or from an incident (which includes, but is not limited to fires) should be added.

[3] Michael J. Karter, Jr. and Joseph L. Molis, *U.S. Firefighter Injuries – 2004.* Fire Analysis and Research Division, National Fire Protection Association, Nov. 2005.

[4] Onduty activities include both fireground and nonfireground operations.

Injuries by Property Type

Eighty-nine percent of firefighter injuries reported to NFIRS in 2004 are associated with structure fires (Figure 3). Of the injuries associated with structure fires, three-quarters (76 percent) occur on residential properties. Overall, structure fires on residential properties account for 68 percent of firefighter injuries. The proportion of firefighter injuries on residential properties has been relatively consistent in the NFIRS 5.0 data. Outside, vehicle, and other fires combined represent 11 percent of firefighter injuries in 2004.

Figure 3. Firefighter Injuries by General Property Type, 2004

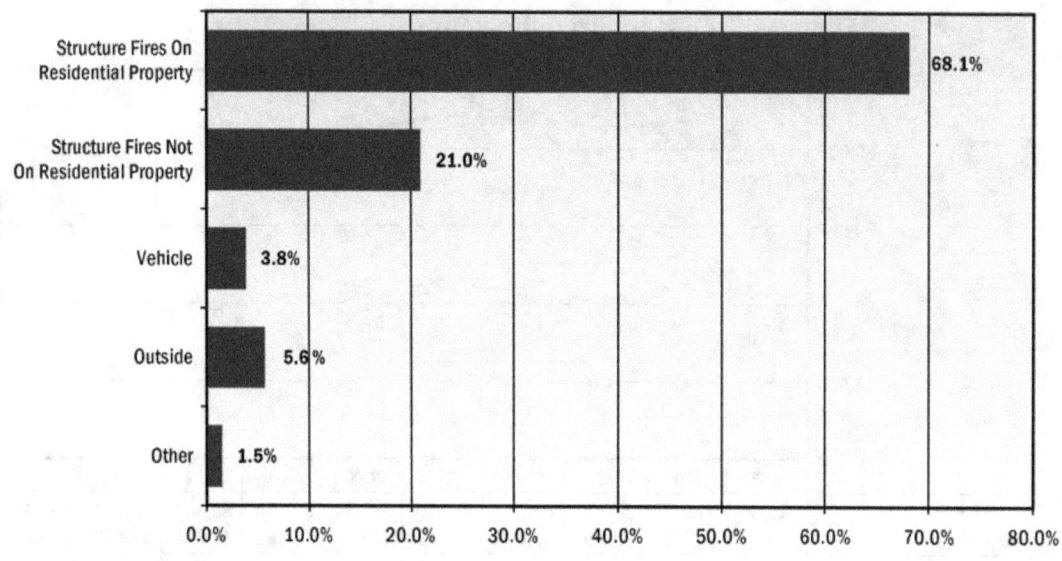

Source: 2004 NFIRS 5.0 data, based on 4,411 firefighter injuries.

Figure 4. Firefighter Injuries by Property Type, Structure Fires Only, 2004

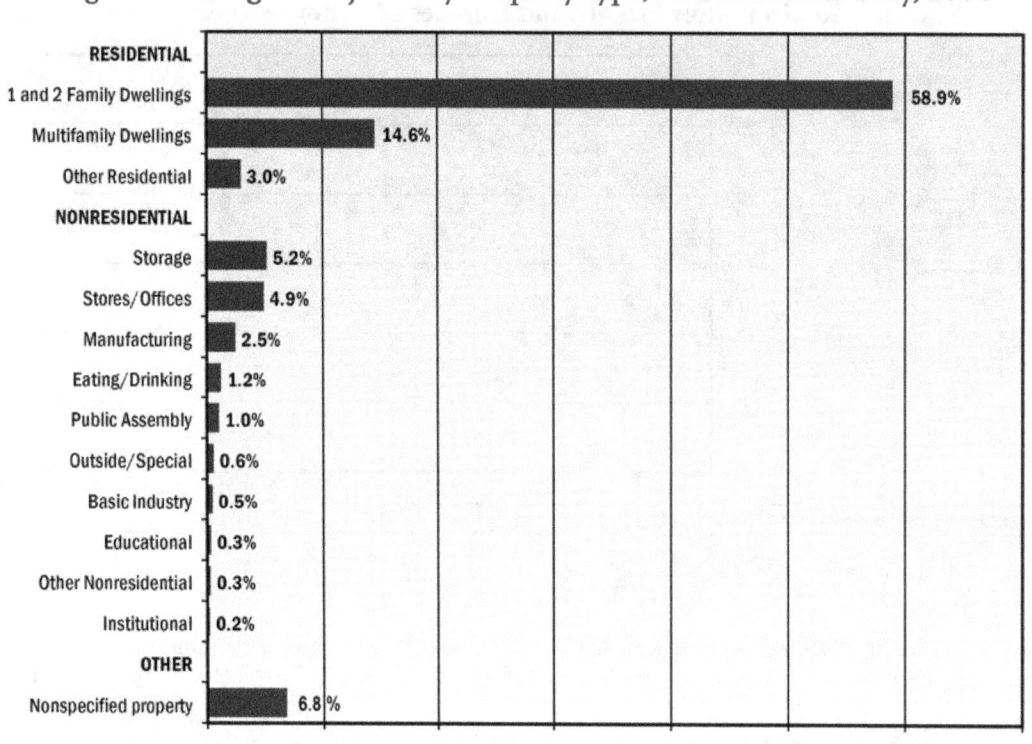

Source: 2004 NFIRS 5.0 data, based on 3,932 structure fire firefighter injuries.

Notes: All structure fire injuries are included, including confined structure fire injuries (107). Property use was not specified for 269 of these injuries.

Figure 4 is a more detailed picture of the relative proportion of structure fire firefighter injuries by property type. Fires on one- and two-family dwelling properties accounted for 59 percent of all firefighter injuries in 2004, a proportion comparable to other years. Multifamily property fires, primarily apartment and row/townhouse fires, accounted for 15 percent.

Figure 4 also shows that two nonresidential property types accounted for the majority of nonresidential structure fire firefighter injuries—storage properties and stores and offices. Stores/Offices long have been a leading nonresidential property type where firefighters have been injured. Storage properties are some of the more challenging properties in which to fight fires, due to the size, layout, and contents of the structure; they often result in substantial injuries. Such an example is the injuries that resulted from an abandoned cold storage warehouse fire in Worcester, MA, in December 1999 in which 399 firefighters were injured.[5, 6]

Injuries per Fire

While injury-producing incidents such as the Worcester, MA, warehouse fire do occur, most fire incidents do not yield injuries. Fire departments focus much of their firefighter training on safety practices and injury prevention, and current protective gear and equipment place a premium on the safety and well-being of the firefighter. When injury-producing incidents do occur, most of them (82 percent) are single-injury incidents (Figure 5). Most injuries (52 percent) do not result in time away from work.

Figure 5. Firefighter Injuries per Injury Incident, 2004

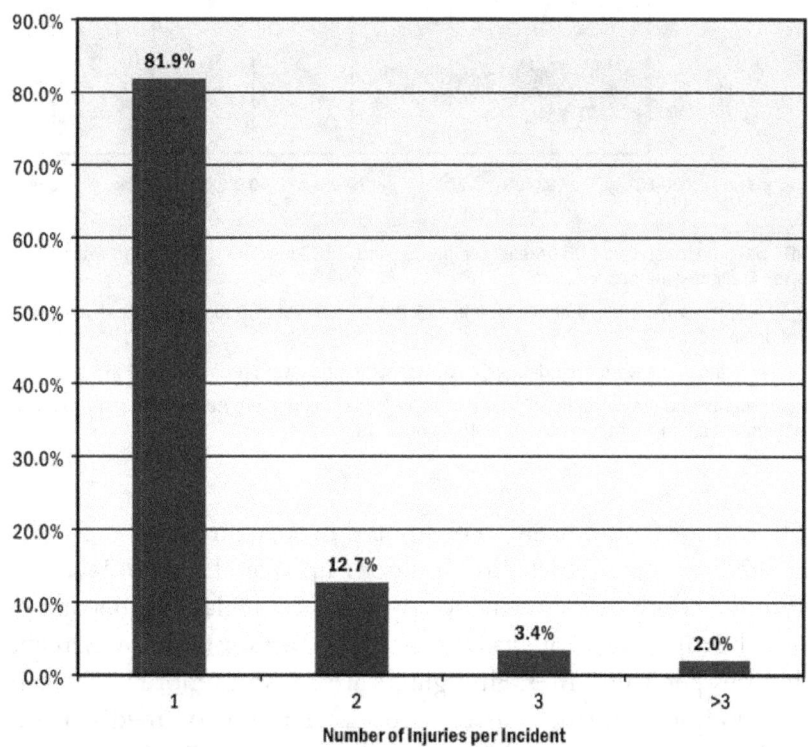

Number of Injuries per Incident

Source: 2004 NFIRS 5.0 data, based on 4,051 firefighter injuries and 3,182 fire incidents.

Notes: New York City structure fire incidents were excluded, as no firefighter injuries were reported to NFIRS in 2004.

Mutual-aid incidents and the corresponding reported firefighter injuries are not included, as insufficient information is available to associate a mutual-aid incident consistently with its originating incident.

---

[5] As reported in the NFIRS database. In this same incident, six firefighters also lost their lives.

[6] In previous USFA analyses of firefighter injuries, vacant and under construction property was a separate property type. Since 1999, firefighter injuries at such sites have been merged with other property types.

While multiple-injury incidents tend to be those where the fire spread is larger (Figure 6), when more firefighters are likely to be involved in the fire operations, and increasing the number of firefighters exposed to potential harm, the injuries generally do not result in lost work time. As discussed in a later section, career firefighters experience proportionally more lost-time injuries than do volunteer firefighters.

Figure 6. Fire Spread for Single- and Multiple-Injury Firefighter Injury Incidents, 2004

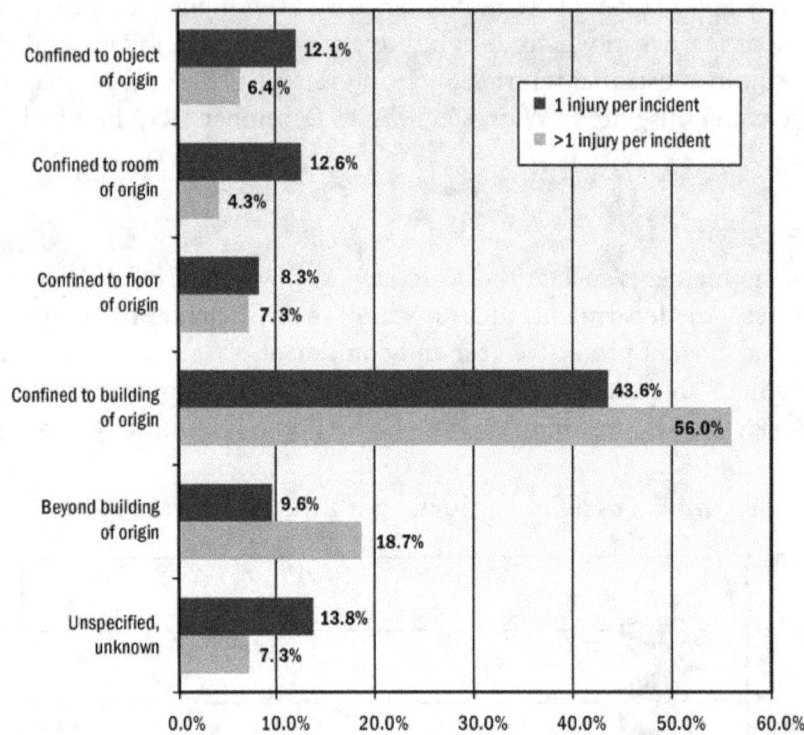

Source: 2004 NFIRS 5.0 data; based on 4,051 firefighter injuries and 3,182 incidents; fire spread was not specified in 491 of the incidents with reported firefighter injuries.

Notes: If fire spread is unknown, and the incident type is 113 to 118, 154, or 155 (confined fires), it is assigned a fire spread of confined to object of origin.

New York City fire incidents were excluded, as no firefighter injuries were reported to NFIRS in 2004.

Mutual-aid incidents and the corresponding reported firefighter injuries are not included, as insufficient information is available to associate a mutual-aid incident consistently with its originating incident.

Firefighters are nearly 15 times more likely to be injured in structure fires, mostly buildings (93 percent), than in nonstructure fires (e.g., vehicle fires, outdoor fires) as shown in Table 1. When confined fires (fires that are confined to noncombustible containers with no flame damage beyond the noncombustible container[7]) are removed, the injury rate for structure fires increases by over 60 percent—from 15 injuries per 1,000 fires to 25 injuries per 1,000 fires. Firefighters are 25 times more likely to be injured in nonconfined structure fires as in nonconfined, nonstructure fires. For nonconfined structure fires, residential structure fires have a higher injury rate per 1,000 fires (27 injuries per 1,000 fires) than nonresidential structures (18 injuries per 1,000 fires). Few firefighters are injured when dealing with confined fires.

---

[7]The definition of confined fires is from the NFIRS 5.0 *Complete Reference Guide* (CRG), Chapter 1, Jan. 2006.

Table 1. Firefighter Injury Rates by Incident Type, 2004

Incident Type	Injuries per 1,000 fires		
	All Fires	Nonconfined Fires	Confined Fires[8]
Structure	**15.4**	**25.0**	**1.1**
Residential	16.5	27.3	1.2
Nonresidential	12.0	18.2	0.9
Nonstructure	**1.0**	**1.0**	**0.6**
Vehicle	1.3	1.3	0.0
Outside and other	0.9	0.9	0.6
Total/Overall	**6.2**	**7.3**	**1.0**

Source: 2004 NFIRS 5.0 data, based on 4,051 firefighter injuries and 656,229 fires.

Notes: New York City fire incidents were excluded, as no firefighter injuries were reported to NFIRS in 2004.

Mutual-aid incidents and the corresponding reported firefighter injuries are not included, as insufficient information is available to associate a mutual-aid incident consistently with its originating incident.

Figure 7. Firefighter Injury Rate per 1,000 Fires by Property Type, Structure Fires, 2004

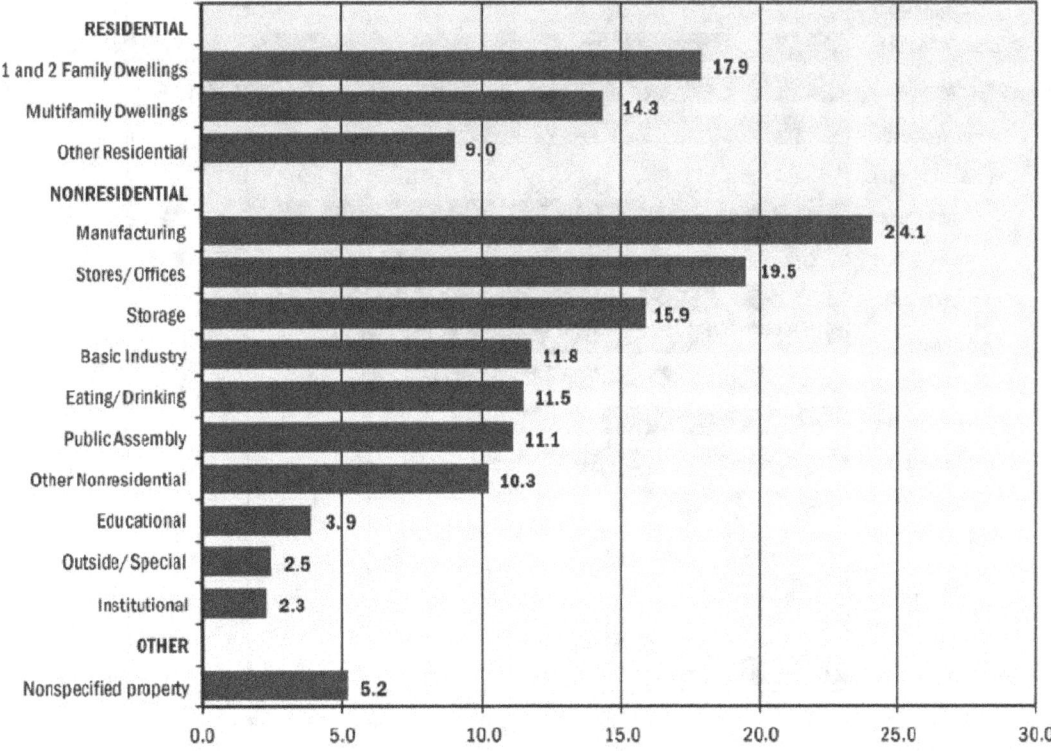

Source: 2004 NFIRS 5.0 data, based on 3,623 structure fire firefighter injuries and 234,873 structure fire incidents.

Notes: All structure fires are included, including confined structure fires.

New York City structure fire incidents were excluded, as no firefighter injuries were reported to NFIRS in 2004.

Mutual-aid incidents and the corresponding reported firefighter injuries are not included, as insufficient information is available to associate a mutual-aid incident consistently with its originating incident.

---

[8] Structure confined fires consist of incident types 113 to 118 (cooking, chimney or flue fire, incinerator, fuel burner/boiler, compactor, or trash/rubbish fires confined to a noncombustible container). Nonstructure confined fires consist of incident types 154 and 155 (dumpster or other outside trash receptacle fire and outside stationary compactor/compacted trash fire).

All residential property types have less risk of firefighter injury per 1,000 fires than manufacturing and commercial properties (Figure 7). In the residential category, one- and two-family residences had the highest injury rate (18 per 1,000 fires). In the nonresidential category, manufacturing properties had the highest injury rate of any property type with 24 firefighter injuries per 1,000 fires. Firefighter injuries in the top three nonresidential properties shown in Figure 7 average 18 injuries per 1,000 fires.

Note that the per-fire injury rates for structure fires in Figure 7 include confined fires and, as Table 1 shows, few firefighters are injured in small, confined fires. As a result, the rates are lower than they would be for more traditional, nonconfined, fires. In Figure 8, confined fires are removed from the analysis and injury rates for nonconfined fires are computed. The injury rates are substantially higher, as expected. The injury risk pattern for nonconfined, nonresidential properties is very similar to that of all structure fires, with some minor reordering in property types. For residential properties, however, a substantially different risk pattern emerges. At 33 injuries per 1,000 fires, multifamily dwellings with nonconfined fires are the riskiest residential property and the riskiest property of all nonconfined fire structure types.

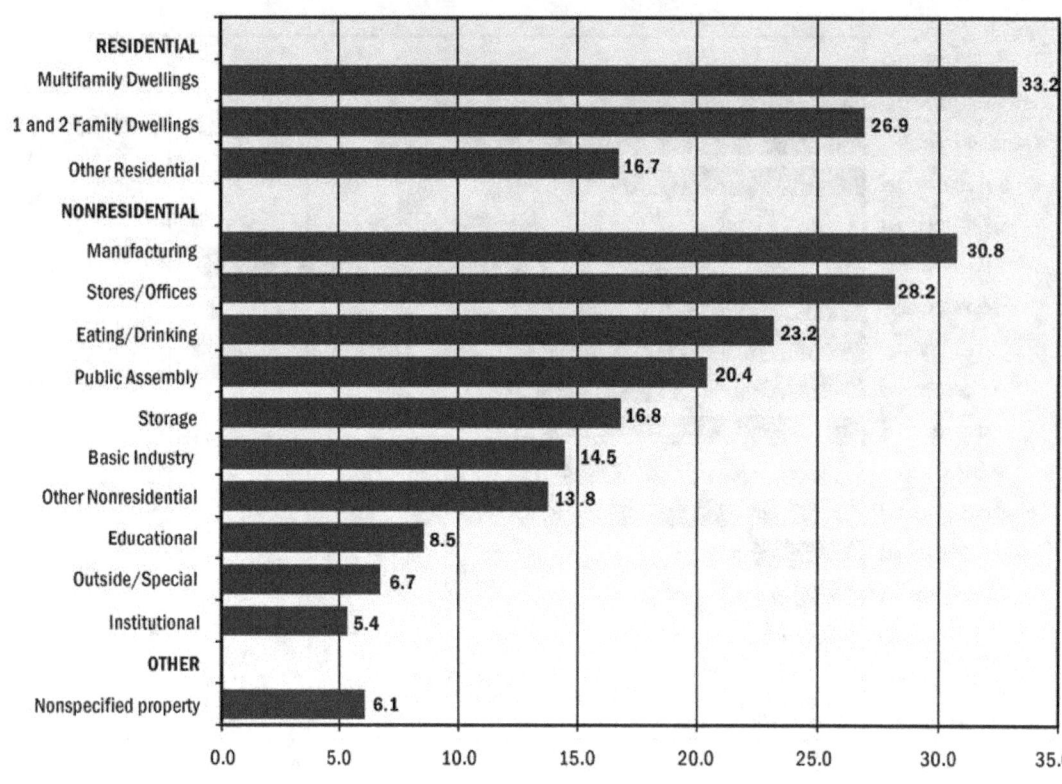

Figure 8. Firefighter Injury Rate per 1,000 Fires by Property Type,
Nonconfined Structure Fires, 2004

Source: 2004 NFIRS 5.0 data, based on 3,519 nonconfined structure fire firefighter injuries and 140,738 nonconfined structure fire incidents.

Notes: New York City fire incidents were excluded, as no firefighter injuries were reported to NFIRS in 2004.

Mutual-aid incidents and the corresponding reported firefighter injuries are not included, as insufficient information is available to associate a mutual-aid incident consistently with its originating incident.

Vacant Properties

Figure 9 shows the rate of firefighter injuries in buildings that are vacant or under construction.[9] Although the injury sample size is relatively small for vacant or under construction properties, the figure

[9]Vacant and under construction buildings are defined by the following criteria: incident types between 111 and 123; structure type null, 1, or 2; building status 1, 4, 5, 6, or 7; mutual-aid incidents are excluded.

shows that 69 firefighters are injured per 1,000 fires in multifamily properties. In years past, nonresidential vacant and under construction properties held the highest risk for firefighter injuries, due in large part to the size of nonresidential buildings. Nevertheless, firefighting challenges also are inherent in multifamily properties. The type of building or complex style (highrise, row/townhouse, midrise) and the number of firefighters often needed to combat the fire can pose significant problems for fire operations.

Vacant and under construction properties have long been a firefighting concern as high-risk sites. The most dangerous fires often are those in vacant properties and properties under construction. In general, these fires are frequently arson-related (51 percent of known causes in 2004), with multiple ignition points. The layout often is unfamiliar and, for properties under construction, continually changing from week to week. Fire defenses built into such structures often are not working or are working only partially. In addition, construction equipment, materials, and debris scattered about the site increase the risk of serious injury. Many of these fires are started when no one is around and the fire spreads rapidly before the fire department is called. This combination continues to make these properties hazardous—in 2004, the overall injury rate for vacant and under construction properties was 33 firefighters per 1,000 fires. This injury rate made these properties one of the most dangerous for firefighters in 2004. As a result of the experience at the Worcester warehouse fire and others like it, today there is less of an inclination to risk firefighters' lives when fighting fires in vacant properties.

Figure 9. Firefighter Injury Rate per 1,000 Fires by Property Type,
Vacant and Under Construction Buildings, 2004

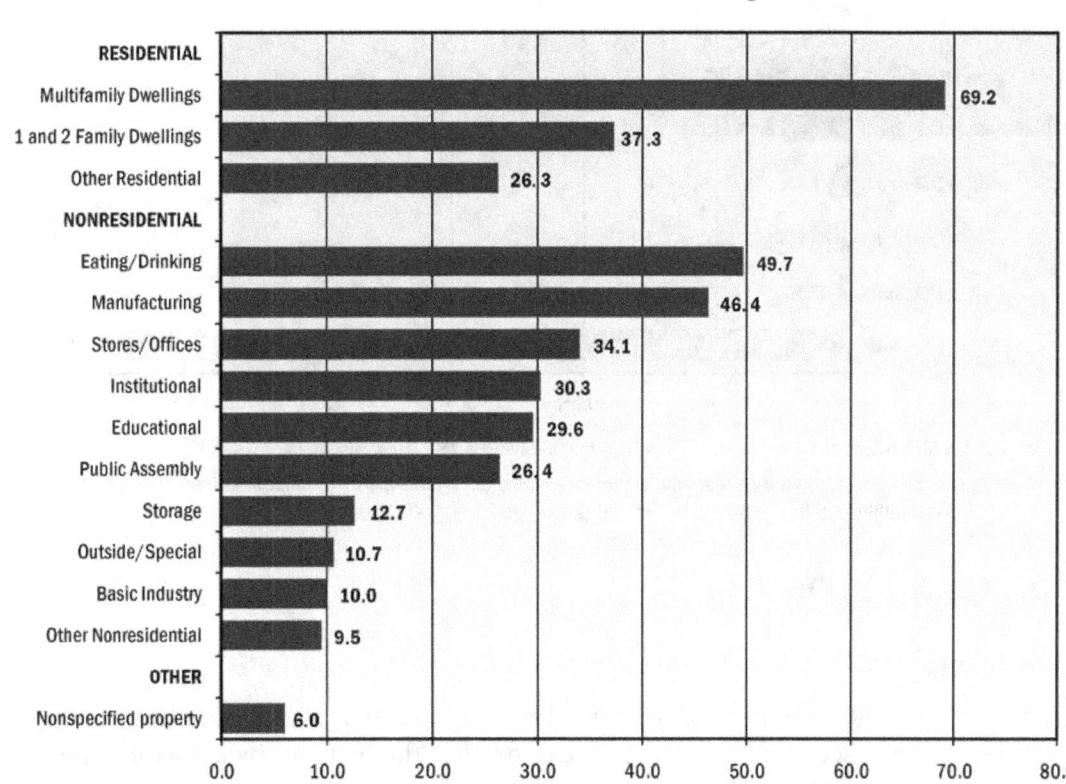

Source: 2004 NFIRS 5.0 data, based on 14,490 vacant and under construction building fire incidents and 477 firefighter injuries.

Notes: New York City fire incidents were excluded, as no firefighter injuries were reported to NFIRS in 2004.

Mutual-aid incidents and the corresponding reported firefighter injuries are not included, as insufficient information is available to associate a mutual-aid incident consistently with its originating incident.

Cause of Injury-Producing Fires

Fifty-five percent of fires that result in firefighter injuries do not have enough information provided to assign a cause, as shown in Figure 10. While this high percentage is startling, what is more problematic is that 69 percent of these "unknown cause" incidents occur on residential properties. Fire prevention for civilians places a premium on determining the cause of fires that injure (and kill) civilians—while the percent of fires with unknown cause is also quite high for civilian injuries (44 percent), relatively speaking, it is 21 percent lower than the 55 percent for firefighter injuries. The same premium on determining fire cause needs to apply to those fires that injure firefighters.

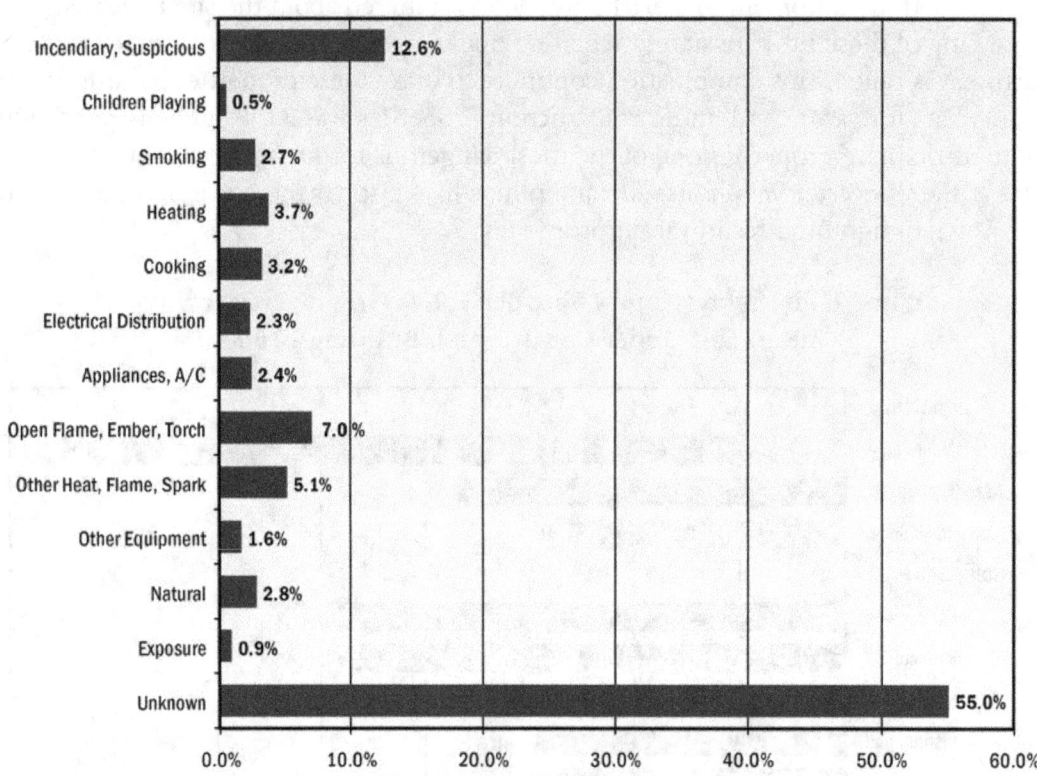

Figure 10. Cause of Injury-Producing Fires for Firefighter Injuries, 2004

Source: 2004 NFIRS 5.0 data; based on 4,051 firefighter injuries and 3,182 incidents.

Note: Mutual-aid incidents and the corresponding reported firefighter injuries are not included, as insufficient information is available to associate a mutual-aid incident with its originating incident and fire cause.

Injury Characteristics

Severity of Injury

More than half of firefighter injuries (52 percent) result in no lost work time as shown in Figure 11. These injuries are treated onscene with first aid, or treated after the incident by a doctor either at a medical facility or in a doctor's office. About 30 percent of firefighter injuries result in lost work time with the bulk of these injuries (29 percent) moderate in severity. Less than 2 percent of injuries are severe or life-threatening. The severity of the injury was not specified for 18 percent of the reported firefighter injuries.

Age

Figure 12 shows the profile of firefighter injuries by age. More than one-third of all injuries (35 percent) occurred to firefighters aged 30 to 39. Firefighter injuries, in general, track the percentage of firefighters in each age group—the more firefighters there are in an age group, the more injuries there are (Figure 13). The resulting injury rates (Figure 14) reveal that firefighters ages 20 to 39 have injury rates 16 to 20 percent higher than the overall injury rate of 33.5 injuries per thousand firefighters.

The types of injuries incurred by firefighters vary with age. Typically, the leading causes of injury among younger firefighters relate to smoke inhalation and exhaustion, and among older firefighters, strains and sprains are the more common injuries. These results relate to physical fitness variations with age, to the effect of age on assignments, and perhaps to the bravado of younger firefighters.

Figure 11. Firefighter Injury Type (Severity of Injury), 2004

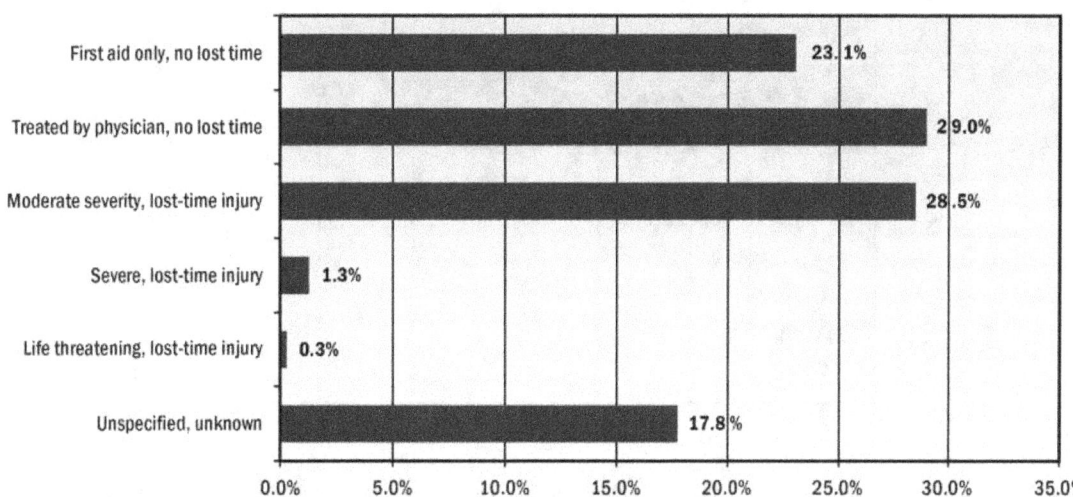

Source: 2004 NFIRS 5.0 data; based on 4,411 firefighter injuries; severity of injury was not specified in 784 of the reported firefighter injuries.

Figure 12. Firefighter Injuries by Age, 2004

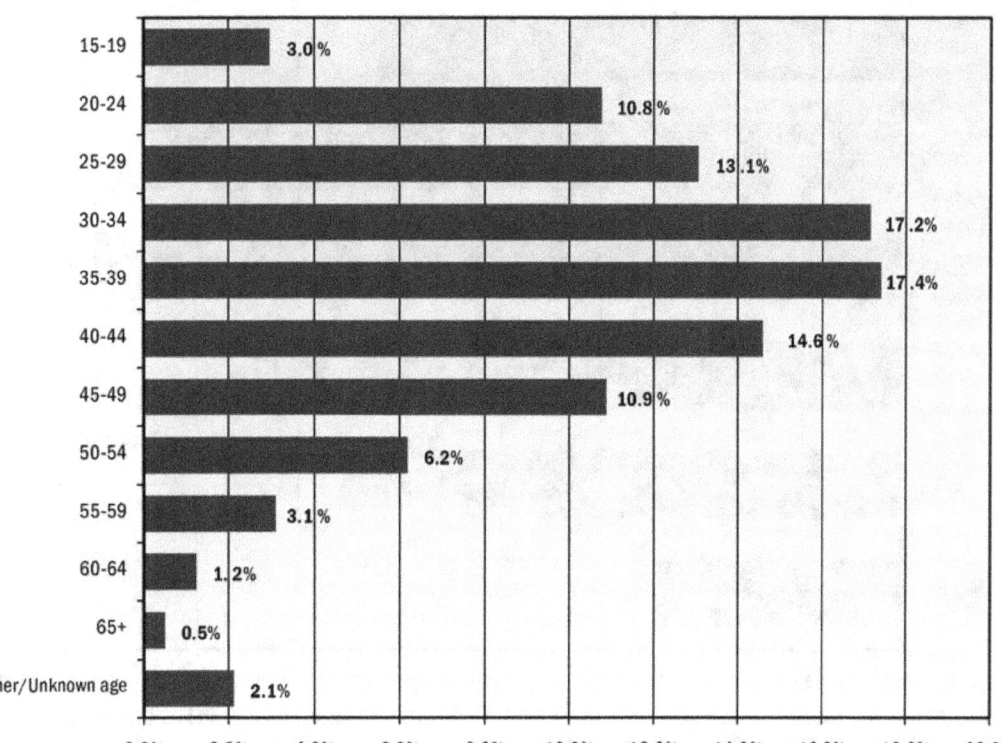

Source: 2004 NFIRS 5.0 data; based on 4,411 firefighter injuries; a valid age was not specified in 94 of the reported firefighter injuries.

Figure 13. Firefighter Injuries and Firefighter Population, 2004

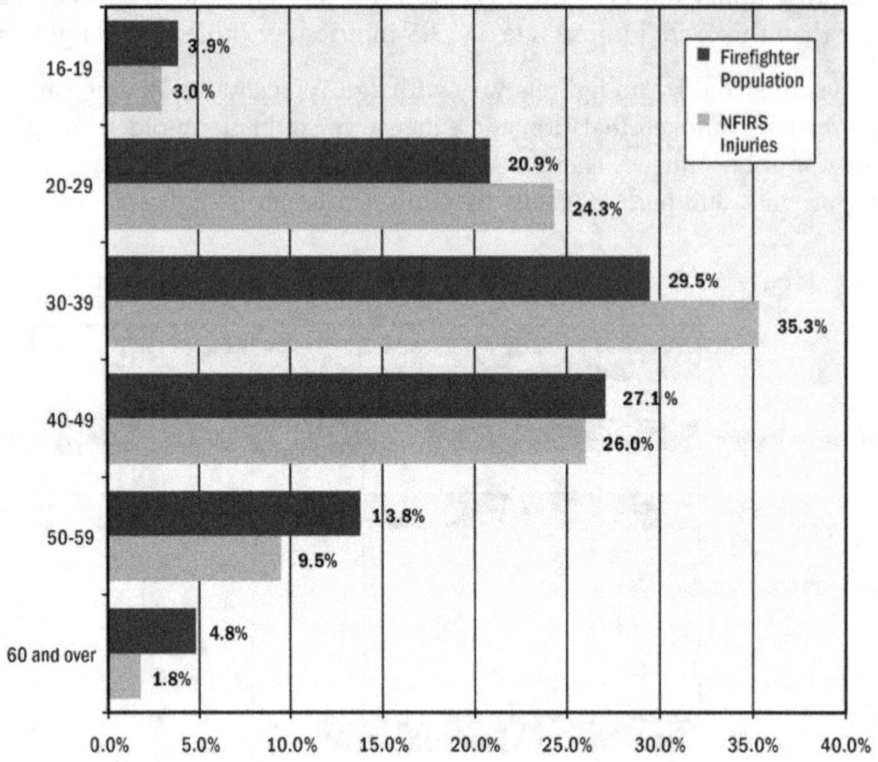

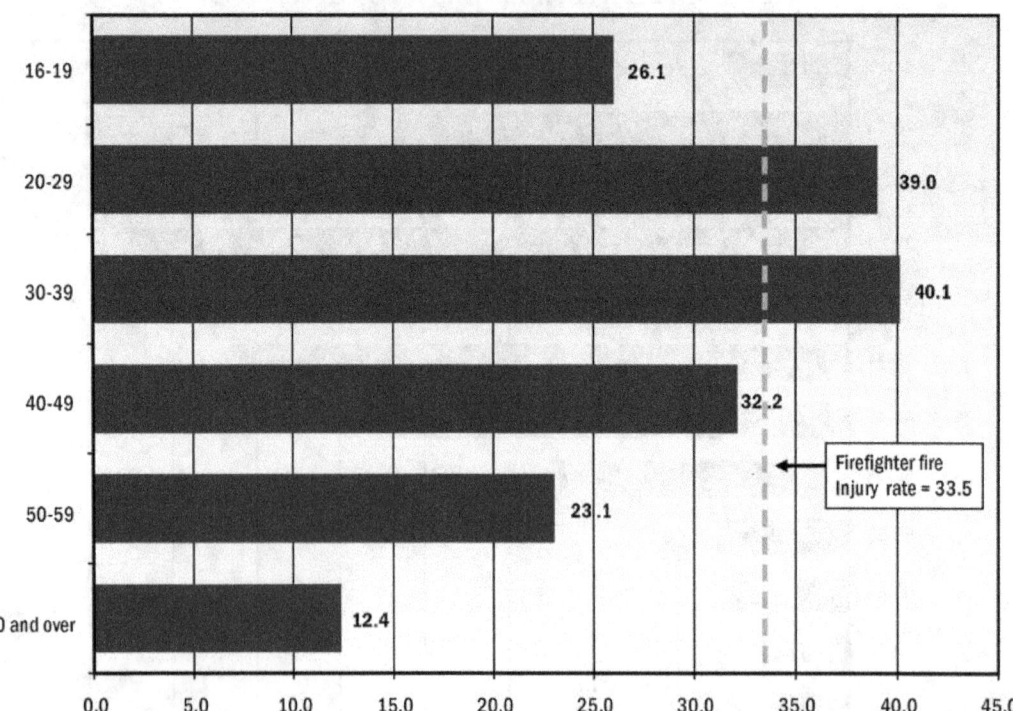

Legend:
- Firefighter Population
- NFIRS Injuries

Age Group	Firefighter Population	NFIRS Injuries
16-19	3.9%	3.0%
20-29	20.9%	24.3%
30-39	29.5%	35.3%
40-49	27.1%	26.0%
50-59	13.8%	9.5%
60 and over	4.8%	1.8%

Source: 2004 NFIRS 5.0 data; based on 4,317 firefighter injuries; a valid age was not specified in 94 of the reported firefighter injuries; *U.S. Fire Department Profile Through 2004*, NFPA.

Figure 14. Firefighter Injury Rates per 1,000 Firefighters by Age Group, 2004

Age Group	Rate
16-19	26.1
20-29	39.0
30-39	40.1
40-49	32.2
50-59	23.1
60 and over	12.4

Firefighter fire Injury rate = 33.5

Source: 2004 NFIRS 5.0 data; based on 4,317 firefighter injuries; a valid age was not specified in 94 of the reported firefighter injuries; *U.S. Fire Department Profile Through 2004*, NFPA.

Gender has only a small effect on the overall age profile of firefighter injuries (Figure 15), with the peak ages for injuries between 35 and 39 for both men and women. Female firefighter injuries occur more often in younger firefighters, peak at ages 35 to 39, then tail off substantially. Male firefighter injuries follow a nearly symmetrical curve. They steadily increase by 3 to 5 percentage points for each age group, reach their highest point between the ages of 30 and 39, and steadily decrease by 3 to 5 percentage points until retirement age.

Figure 15. Firefighter Injuries by Age and Gender, 2004

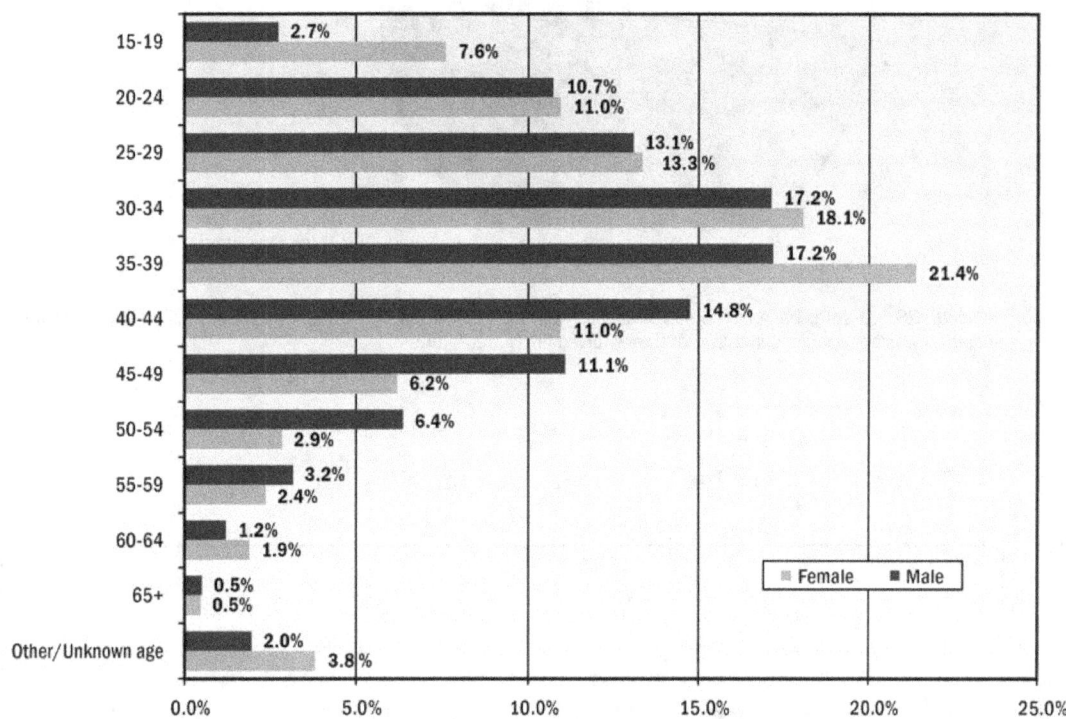

Source: 2004 NFIRS 5.0 data; based on 4,121 male firefighter injuries and 210 female firefighter injuries; gender was not specified in 80 of the reported firefighter injuries.

Affiliation

Injuries to career firefighters are the largest share (44 percent) of the reported injuries (Figure 16). This is in contrast to firefighter deaths—in 2004, 70 percent of firefighter deaths were from the volunteer fire service.[10] As shown in Figure 17, injuries to career firefighters tend to occur in midcareer (ages 30 to 45) with the peak between ages 35 and 39. Injuries to volunteers, on the other hand, are sustained predominately by the younger members of the organization. Firefighters under the age of 25 account for nearly 28 percent of injuries in the volunteer service.

---

[10] *Firefighter Fatalities in the United States in 2004.* U.S. Department of Homeland Security, Federal Emergency Management Agency, U.S. Fire Administration, Aug. 2005.

Figure 16. Firefighter Injuries by Affiliation, 2004

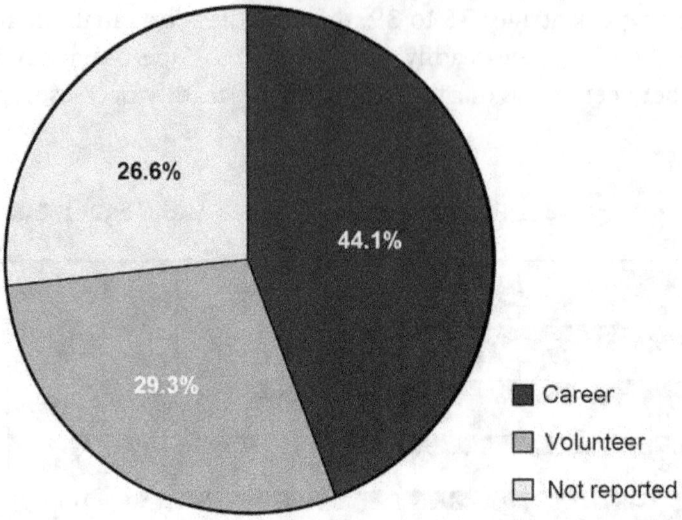

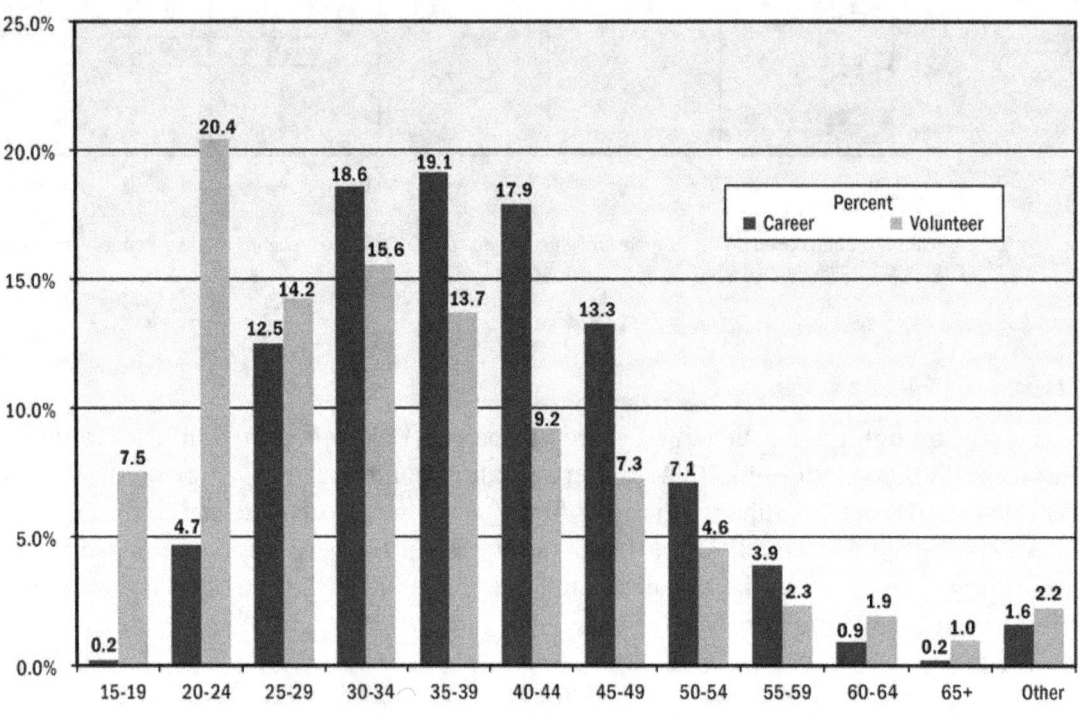

Source: 2004 NFIRS 5.0 data; based on 4,411 firefighter injuries (1,944 career firefighter injuries; 1,292 volunteer firefighter injuries; affiliation was not specified in 1,175 of the reported firefighter injuries).

Figure 17. Career and Volunteer Firefighter Injuries by Age, 2004

Source: 2004 NFIRS 5.0 data; based on total of 3,236 firefighter injuries (1,944 career firefighter injuries and 1,292 volunteer firefighter injuries) where affiliation was specified.

Career firefighters also experience proportionally more lost-time injuries than their volunteer counter-parts (approximately 2 to 1) as shown in Table 2 below.

Table 2. Overall Comparison of Injury Severity by Affiliation, 2004, Percent

Affiliation	Severity			Total
	No Lost-time	Lost-time	Severity Not Specified	
Overall	52.1	30.1	17.8	100.0
Career	48.4	39.6	12.0	100.0
Volunteer	65.9	18.7	15.5	100.0
Affiliation not specified	43.2	27.0	29.8	100.0

Source: 2004 NFIRS 5.0 data; based on total of 4,411 firefighter injuries (1,944 career firefighter injuries; 1,292 volunteer firefighter injuries; affiliation was not specified in 1,175 of the reported firefighter injuries). Severity was not specified in 784 of the reported firefighter injuries; 350 reported firefighter injuries had neither affiliation nor severity level specified.

Note: Totals may not add due to rounding.

When Injuries Occur

TIME OF DAY. As in previous years, injuries to firefighters peak in the early afternoon and again in the evening and are lowest in the mornings. In 2004, firefighter injuries reached their lowest point between 8 and 9 a.m. and peaked between 2 and 3 p.m. and again between 7 and 8 p.m. Half of all firefighter injuries occurred between 1 p.m. and 11 p.m. (Figure 18). The times that are most hazardous to civilians (evening mealtimes for injuries) are also the times when more firefighters are injured. As shown in Figure 18, the daily cycle of injuries roughly follows that of fires but, as to be expected, is offset by several hours.

Figure 18. Firefighter Injuries by Time of Day, 2004

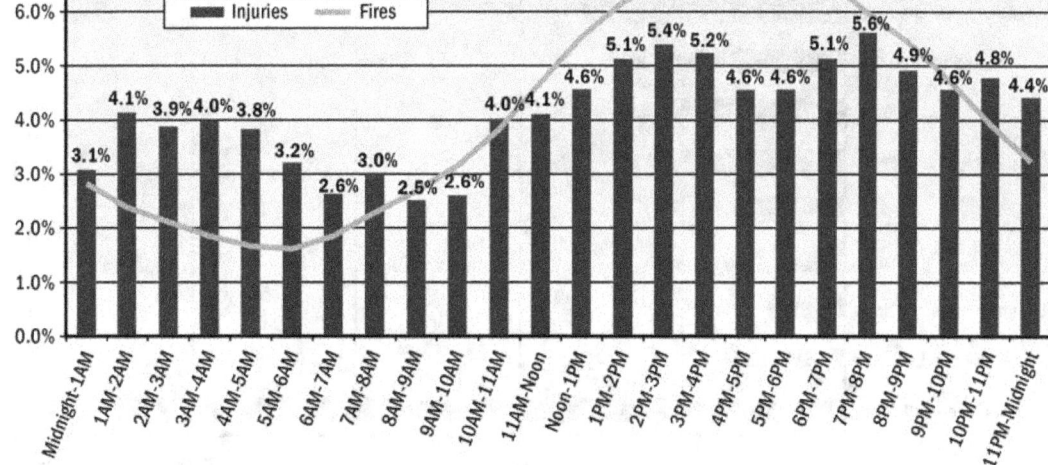

Source: 2004 NFIRS 5.0 data; based on 4,411 firefighter injuries.

MONTH OF YEAR. Firefighter injuries are somewhat higher in midwinter (January–February) when residential fires peak (and conditions are more severe in much of the Nation) (Figure 19). Injuries dip from September through November, reaching their lowest point in November.

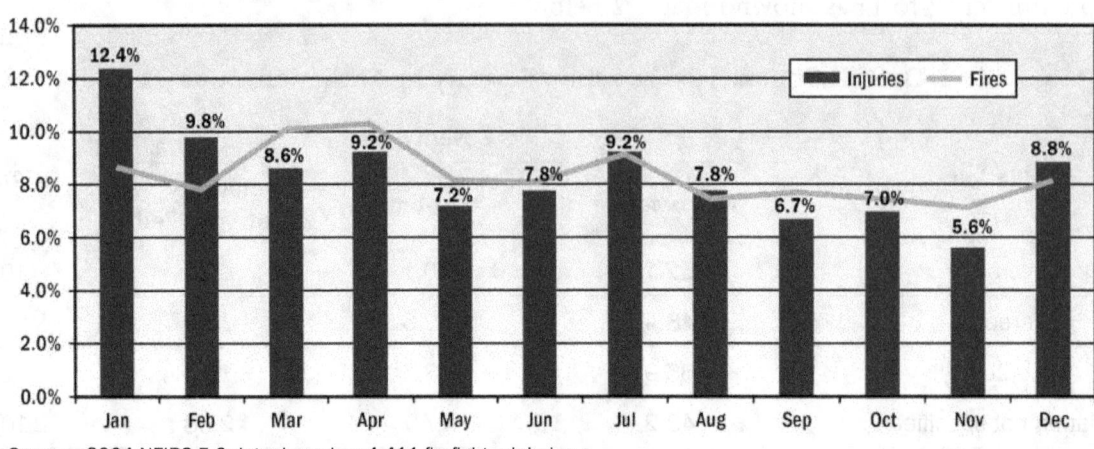

Figure 19. Firefighter Injuries by Month, 2004

Source: 2004 NFIRS 5.0 data; based on 4,411 firefighter injuries.

Part of Body Injured

Thirty percent of firefighter injuries in 2004 were to the upper and lower extremities (torso, arms/hands, and legs/feet) (Figure 20). The head and shoulder areas account for an additional 20 percent of injuries. All areas of the body are vulnerable, however. Where no body part is specified, the primary symptoms are dizziness and exhaustion (3 percent). No injury area was indicated in 22 percent of firefighter injuries.

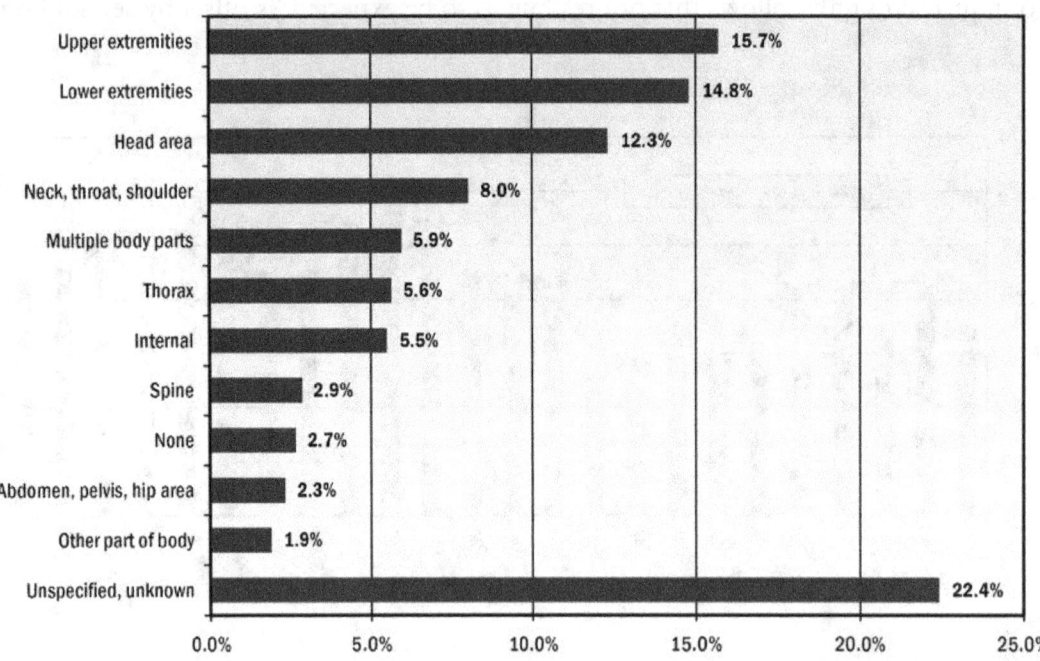

Figure 20. Firefighter Injuries by Part of Body Injured, 2004

Source: 2004 NFIRS 5.0 data; based on 4,411 firefighter injuries; primary part of body injured was not specified in 989 of the reported firefighter injuries.

Cause of Injury

As shown in Figure 21, the greatest cause of firefighter injuries associated with fires was reported to be overexertion and strains (16 percent), followed by contact with or exposure to flames or smoke (16 percent). These two injury causes alone reinforce the fact that firefighting is a physically exhausting and dangerous profession. The cause of firefighter injury was undetermined in 26 percent of cases.

Figure 21. Firefighter Injuries by Cause of Injury, 2004

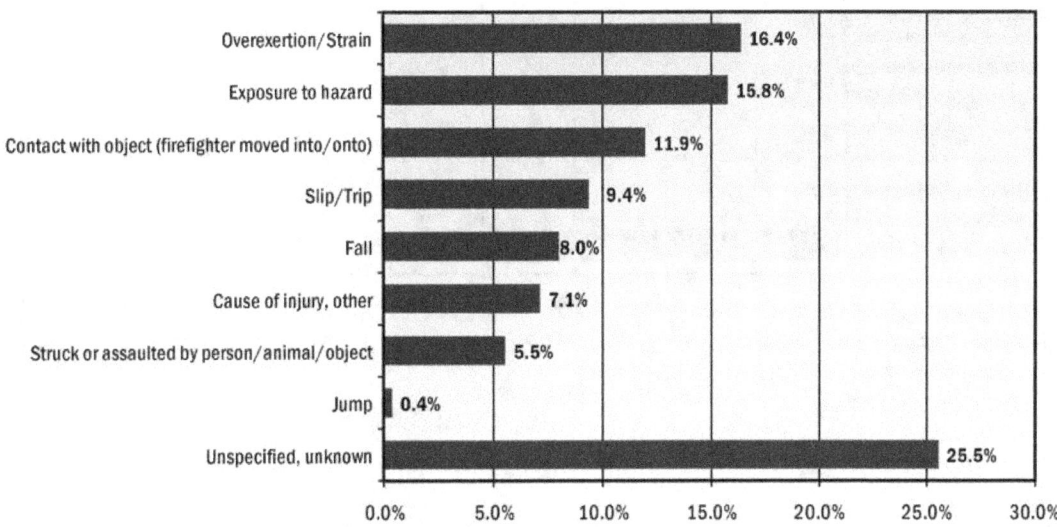

Source: 2004 NFIRS 5.0 data; based on 4,411 firefighter injuries with cause of injury specified; cause of injury was not specified in 1,127 of the reported firefighter injuries.

Where Injuries Occur

Nearly three-quarters (73 percent) of the 2004 firefighter injuries occur at the fire scene (Figure 22). The striking point here is that many firefighter injuries (38 percent) occur in areas outside the fire building, a place where the firefighter may feel relatively safe. There often are more firefighters operating outside the fire building and exposed to injury than there are inside. At scene, outside structure fires include vehicle fires, which contribute to this high incidence of injuries. This distribution has changed little over the years. Where the injury occurred was not specified in 23 percent of the reported firefighter injuries.

Less than 2 percent of firefighter injuries occurred in the fire station and approximately 1 percent occurred while traveling to or from assignments, to the fire department, or to a medical facility for treatment.

Factor Contributing to Injury

When a factor was specified as contributing to the firefighter's injury, fire development—fire progress, smoking conditions, and the like—and slippery and uneven surfaces account for 29 percent of firefighter injuries, with fire development as the leading factor contributing to injury. Structures and outside and other incidents have a higher proportion of injuries resulting from fire development. Vehicle incidents have a higher proportion of injuries resulting from surface issues, ostensibly wet road surfaces due to water or foam, as the major contributing factor.

Figure 22. Firefighter Injuries by Where Injury Occurred, 2004

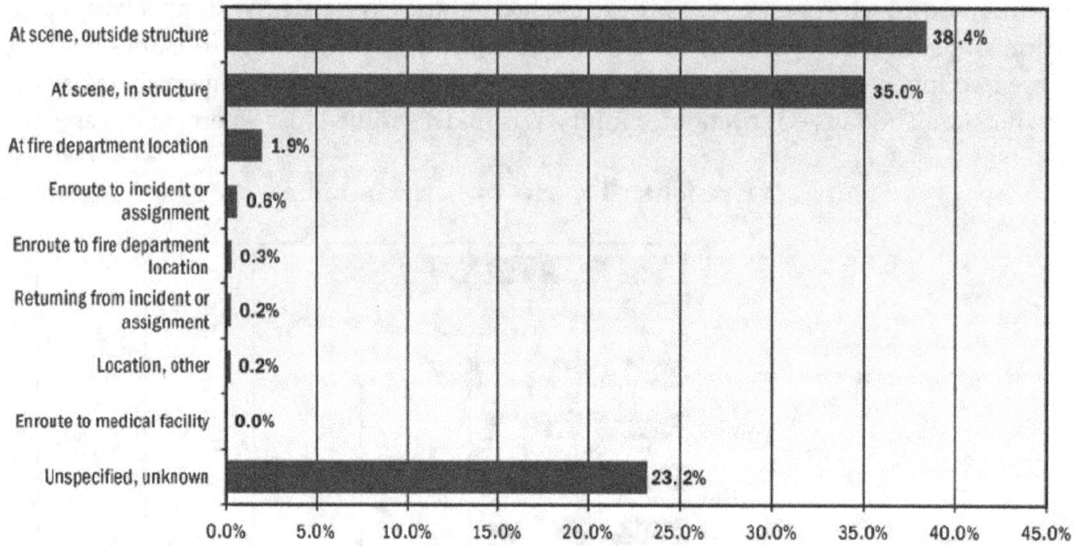

Source: 2004 NFIRS 5.0 data; based on 4,411 firefighter injuries; where injury occurred was not specified in 1,024 of the reported firefighter injuries.

Table 3. General Factor Contributing to Firefighter Injuries, 2004, Percent

General Factor Contributing to Injury	Incident Type			
	Overall	Structures	Vehicle	Outside and Other
Fire development	**16.0**	**16.2**	13.2	**15.1**
Slippery or uneven surfaces	12.8	12.6	**15.6**	14.1
Collapse or falling object	7.8	8.2	3.0	4.2
Holes	1.8	1.9	0.6	1.3
Vehicle or apparatus issue	1.7	1.2	4.8	5.8
Lost, caught, trapped, or confined	1.4	1.6	0.6	0.0
Civil unrest/Hostile acts	0.5	0.5	0.0	0.6
Other factor	12.0	11.4	12.6	18.6
No factor involved	13.4	13.3	18.0	11.9
Unspecified, unknown	32.8	33.2	31.7	28.5
Total	100.0	100.0	100.0	100.0

Source: 2004 NFIRS 5.0 data; based on 4,411 firefighter injuries. Includes 1,448 reported firefighter injuries where factor contributing to injury was not
specified.

Note: Totals may not add due to rounding.

Protective Equipment Failure

Very few of the firefighter injuries reported to NFIRS indicated problems with firefighter protective gear—only 7 percent indicated protective equipment failures as a factor in the injury. Modern equipment and equipment standards, combined with current equipment replacement cycles, may preclude protective equipment failures. Protective coats and firefighter gloves with wristlets accounted for 25 percent of equipment problems. Protective gear failure was not reported in 8 percent of the reported firefighter injuries.

Responses and Physical Condition Prior to Injury

Most firefighters (over 70 percent) reported being well-rested before their injury—this applies to both minor and severe injuries as shown in Table 4.

Table 4. Firefighter Physical Condition Prior to Injury, 2004, Percent

Physical Condition Prior to Injury	Severity Level			Overall
	No Lost-Time	Lost-Time	Severity Not Specified	
Rested	70.5	72.4	7.7	59.9
Fatigued	9.2	9.1	0.9	7.7
Ill or injured	1.3	2.1	0.3	1.4
Physical condition, other	1.8	1.0	0.3	1.3
Unspecified, unknown	17.2	15.4	90.9	29.8
Total	100.0	100.0	100.0	100.0

Source: 2004 NFIRS 5.0 data; based on 4,411 firefighter injuries. Severity level was not specified in 784 of the reported firefighter injuries; condition prior to injury was not specified in 1,313 of the reported firefighter injuries (713 reported firefighter injuries had neither condition prior to injury nor severity level specified).

Note: Totals may not add due to rounding.

The number of fire department responses attended prior to the injury, however, does result in more severe injuries. Table 5 shows firefighters with one or more prior responses have a higher percentage of lost-time injuries than firefighters who reported no prior responses. Moreover, it is the moderate lost-time injuries that are most affected by prior responses (Figure 23).

Table 5. Responses Prior to Firefighter Injury, All Injuries, 2004, Percent

Number of Responses Prior to Injury	Severity Level			Total
	No Lost-Time	Lost-Time	Severity Not Specified	
No prior responses	50.8	26.3	22.9	100.0
One or more prior responses	55.3	39.4	5.3	100.0

Source: 2004 NFIRS 5.0 data; based on 4,411 firefighter injuries; severity level was not specified in 784 of the reported firefighter injuries.

Figure 23. Severity of Injury by Responses Prior to Firefighter Injury, 2004

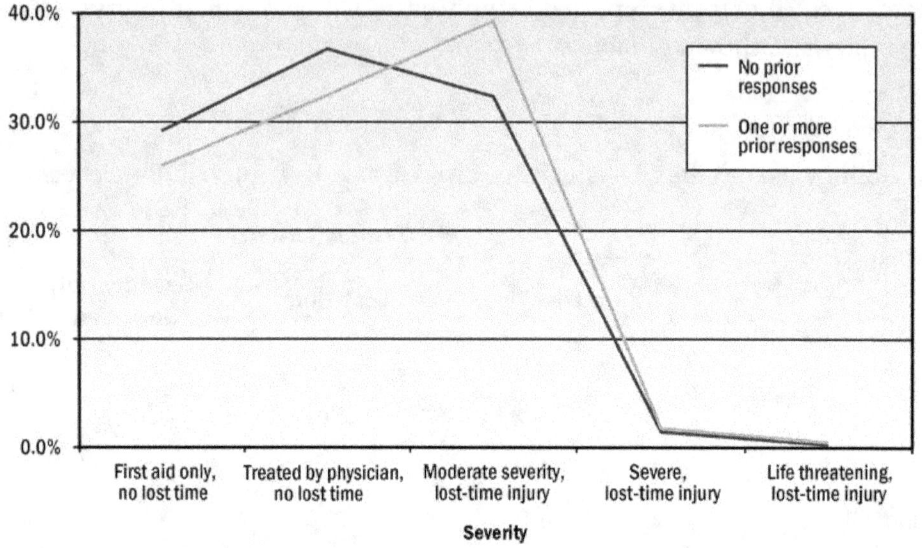

Source: 2004 NFIRS 5.0 data; based on 3,627 firefighter injuries where the severity level was specified. Severity level was not specified in 784 of the reported firefighter injuries.

Type of Activity When Injured

Similar to previous years, 41 percent of firefighter injuries in 2004 occurred while extinguishing the fire; suppression support accounted for 20 percent (Figure 24). The activity when the firefighter was injured was not specified in 20 percent of the reported firefighter injuries.

Figure 24. Firefighter Injuries by Type of Activity, 2004

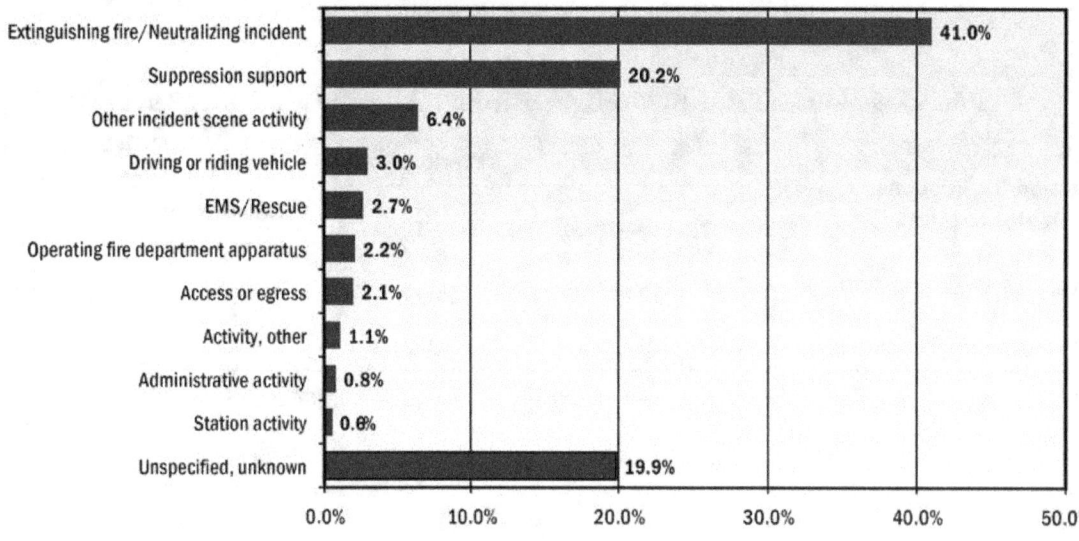

Source: 2004 NFIRS 5.0 data; based on 4,411 firefighter injuries; activity at time of injury was not specified in 876 of the reported firefighter injuries.

Nature of Injury

There is a sharp contrast between the nature of firefighter injuries and the nature of deaths. Heart attacks and internal trauma accounted for the majority of firefighter fatalities in 2004 (79 percent),[11] but these equivalent categories for injuries accounted for only 2 percent of firefighter injuries in 2004 (Figure 25). Sprains and strains and cuts and wounds accounted for over a third of injuries (34 percent). Burns accounted for an additional 11 percent. The type of injury was not reported in 19 percent of firefighter injuries.

Figure 25. Firefighter Injuries by Nature of Injury, 2004

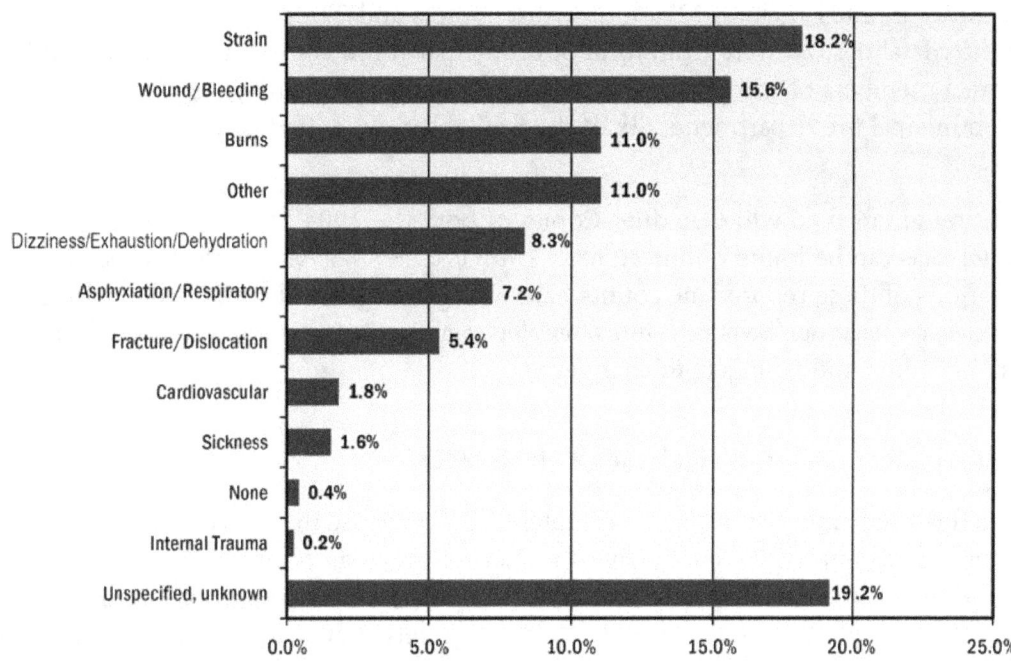

Source: 2004 NFIRS 5.0 data; based on 4,411 firefighter injuries; nature of injury was not specified in 845 of the reported firefighter injuries.

Type of Medical Care

Forty-nine percent of the reported firefighter injuries associated with fires in 2004 were treated at hospitals (Figure 26). Another 24 percent were treated but not transported. Very few firefighters seek medical care at a doctor's office. For 20 percent of injured firefighters, the treatment location was not specified.

Figure 26. Firefighter Injuries by Where Treated, 2004

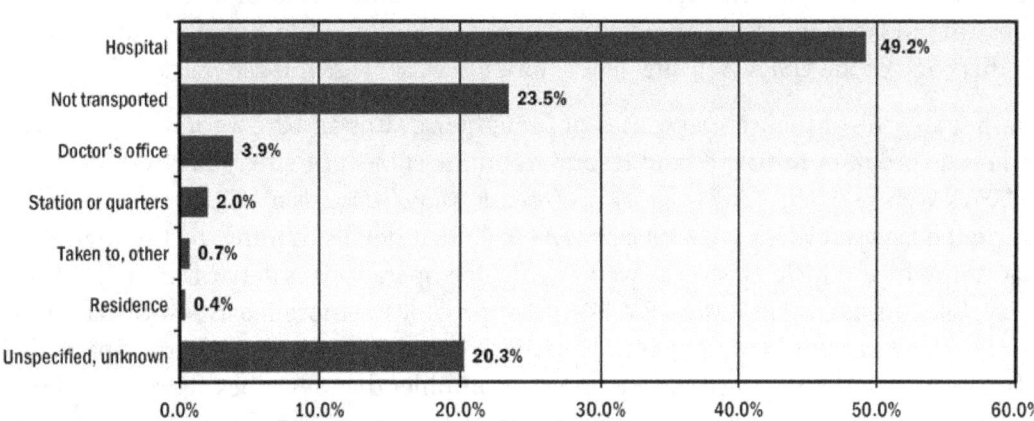

Source: 2004 NFIRS 5.0 data; based on 4,411 firefighter injuries; treatment location was not specified in 895 of the reported firefighter injuries.

[11] *Firefighter Fatalities in the United States in 2004.* U.S. Department of Homeland Security, Federal Emergency Management Agency, U.S. Fire Administration, Aug. 2005.

Firefighter Deaths

Each year, USFA produces studies of onduty firefighter fatalities in the United States and its protectorates. One hundred and nineteen firefighters died while on duty in 2004. In 2004, this total was affected by a change in the inclusion criteria—there were 11 firefighter fatalities that would not have been considered for inclusion prior to the enactment of the Hometown Heroes Survivors Benefit Act of 2003. This law generated a change in report criteria to include firefighters who die within 24 hours of stressful onduty activity.

Firefighter fatalities in 2004 include 82 volunteer firefighters and 37 career firefighters. Among the volunteer firefighter fatalities, 73 were from local or municipal volunteer fire departments, and 9 were seasonal or contract members of wildland fire agencies. All of the career firefighters who died were members of local or municipal fire departments. Six of the firefighters who died in 2004 were female, and 113 were male.

In 2005, 115 firefighters died while on duty. Copies of both the 2004 and the 2005 reports on *Firefighter Fatalities in the United States* can be found online at: http://www.usfa.dhs.gov/fireservice/fatalities/statistics/report.shtm Since the publications of these reports, the counts have changed slightly. The updated counts can be found at: http://www.usfa.dhs.gov/fireservice/fatalities/statistics/casualties.shtm For 2006, there were 106 firefighter deaths. This is a provisional count and is subject to change.

USFA Resources on Firefighter Injuries

USFA-supported research and development are intended to increase the safety and well-being of emergency response personnel. USFA encourages the sharing of research findings and incorporation of innovations in equipment available to firefighters and other responders through programs that focus on health and safety studies; research, training, and awareness; emergency medical services (EMS); search and rescue; and equipment and technology development.

As heart attacks and stress are the leading causes of onduty firefighter fatalities and are the primary source of life-threatening injuries, USFA has developed materials addressing firefighter health and wellness programs to mitigate this. Among these are the *Health and Wellness Guide for the Volunteer Fire Service* (FA-267), developed in partnership with the National Volunteer Fire Council (NVFC), and the *Emergency Incident Rehabilitation* (FA-114) booklet, which USFA currently is working to update with the International Association of Fire Fighters (IAFF). This booklet includes a sample standard operating procedure (SOP) and guidelines for establishing a rehab area to reduce heat- or cold-related injuries to emergency response personnel operating in labor-intensive or extreme climate conditions. Additional initiatives in this area are described on this page of the USFA Web site: http://www.usfa.dhs.gov/research/safety/fitness.shtm

Because crashes are one of the leading causes of firefighter deaths and are a source of firefighter injuries, USFA has numerous program initiatives and resources on the subject of emergency vehicle safety; these are detailed on USFA's Web site: http://www.usfa.dhs.gov/research/safety/vehicle.shtm USFA publications of interest in this area include the *Emergency Vehicle Safety Initiative* (FA–272) that details training, technological, and other programs that can reduce vehicle crashes as well as enhance operational safety of firefighters operating on the roadway; *Safe Operations of Fire Tankers* (FA–248) that provides comprehensive information regarding the safety practices and principals of operating fire tanker vehicles for local-level fire departments; *Alive on Arrival—Tips for Safe Emergency Vehicle Operations* (FA–255), a pamphlet that describes actions that emergency vehicle operators, passengers, and officers-in-charge can take to improve safety; and *Emergency Vehicle Driver Training* (FA–110), a training package that includes both an instructor manual and a student workbook. Also available is a special report titled *Fire Apparatus/Train Collision* (USFA–TR–048) that presents the investigation of the collision near Catlett, VA, on September 28, 1989.

As a follow on to the *Emergency Vehicle Safety Initiative*, USFA has formed partnerships with several leading fire service membership organizations to develop comprehensive Web-based emergency vehicle safety educational programs targeted for their specific constituents. The following educational programs have been developed in conjunction with the USFA:

Guide to Model Policies and Procedures for Emergency Vehicle Safety, International Association of Fire Chiefs, (IAFC).

Improving Apparatus Response and Roadway Operations Safety in the Career Fire Service, IAFF.

Emergency Vehicle Safe Operations for Volunteer and Small Combination Emergency Service Organizations, NVFC.

In addition, USFA has several project and partnership efforts aimed to reduce firefighter and emergency responder onduty injuries and deaths resulting from being struck while operating on the roadway: http://www.usfa.dhs.gov/research/safety/roadway.shtm

USFA is currently revising the *Firefighter Autopsy Protocol* (May 1995) (FA-156), to provide the most current and comprehensive, step-by-step protocol for performing autopsies for firefighters and emergency medical service responders killed in the line of duty.

The USFA has developed a publication for emergency response managers on infection control programs based on Federal laws, regulations, and standards. The *Guide to Developing and Managing an Emergency Service Infection Control Program* (FA–112) addresses modes of disease transmission, measures for prevention, incident response and recovery, station issues, and training/role modeling. The manual provides a step-by-step approach to designing, implementing, managing, and evaluating a fire or EMS department infection control program. The guide is also a key resource in a National Fire Academy (NFA) course on infection control.

The *Topical Fire Report Series* for firefighter injuries can be downloaded from http://www.usfa.dhs.gov/statistics/reports/tfrs_issue_index.shtm

Firefighter Injuries, Volume 2, Issue 1
Firefighter Injuries in Structures, Volume 2, Issue 2

Reports produced under the USFA's Major Fires Investigation series are directed primarily to chief fire officers, training officers, fire marshals, and investigators as a resource for training and prevention. The Technical Report Series on incidents involving firefighter casualties (deaths and injuries) includes

Abandoned Cold Storage Warehouse Multi-Firefighter Fatality Fire (Worchester, MA-December 1999) (USFA-TR-134)
Aerial Ladder Collapse Incidents (April 1996) (USFA-TR-081)
Confined Space Rescue on SS Gem State, Tacoma, WA (FA-163A)
Detroit Warehouse Fire Claims Three Firefighters (March 1987) (USFA-TR-003)
Entrapment in Garage Kills One Firefighter (San Francisco, CA-March 1995) (USFA-TR-084)
Floor Collapse Claims Two Firefighters (Pittston, PA-March 1993) (USFA-TR-073)
Four Firefighters Die in Seattle Warehouse Fire (Seattle, WA-January 1995) (USFA-TR-077)
Four Firefighters Killed, Trapped by Floor Collapse (Brackenridge, PA-December 1991) (USFA-TR-061)
Highrise Office Building Fire, One Meridian Plaza (Philadelphia, PA-February 1991) (USFA-TR-049)
Indianapolis Athletic Club Fire (Indianapolis, IN-February 1992) (USFA-TR-063)
LP Gas Tank Explosion Kills Two Volunteer Firefighters (Carthage, IL-October 1997) (USFA-TR-120)
Multiple Fatality Highrise Condominium Fire (Clearwater, FL-June 2002) (USFA-TR-148)
National Guard Plane Crash at Hotel Site (Evansville, IN-February 1992) (USFA-TR-064)

Santana Row Development Fire (San Jose, CA-August 2002) (USFA-TR-153)
Six Firefighter Fatalities in Construction Site Explosion (Kansas City, MO-November 1988) (USFA-TR-024)
Sodium Explosion Critically Burns Firefighters (Newton, MA-October 1993) (USFA-TR-075)
Structural Collapse at Dwelling Fire Results in Two Firefighter Fatalities (Stockton, CA-February 1997) (USFA-TR-102)
Three Firefighter Fatalities in Training Exercise (Milford, MI-October 1987) (USFA-TR-015)
Three Firefighters Die in Pittsburgh House Fire (Pittsburgh, PA-February 1995) (USFA-TR-078)
Two Firefighter Deaths in Auto Parts Store Fire (Chesapeake, VA-March 1996) (USFA-TR-087)
Wood Truss Roof Collapse Claims Two Firefighters (Memphis, TN-December 1992) (USFA-TR-069)

Other USFA works of interest regarding firefighter casualties:

~ A Heat Transfer Model for Fire Fighter's Protective Clothing (January 1999) (FA-192)
~ A Needs Assessment of the U.S. Fire Service (December 2002) (FA-240)
  Aftermath of Firefighter Fatality Incidents: Preparing for the Worst: Special Report (USFA-TR-089)
~ America at Risk (June 2002) (FA-223)
~ Developing Effective Standard Operating Procedures for Fire & EMS Departments (December 1999) (FA-197)
  Emergency Medical Services (EMS) Recruitment & Retention Manual (FA-157)
~ EMS Safety Techniques and Applications (April 1994) (FA-144)
  Fire and Emergency Medical Services Ergonomics Manual — A Guide for Understanding and Implementing An Ergonomics
      Program in Your Department (FA-161)
~ Fire Department Communications Manual: A Basic Guide to System Concepts and Equipment (FA-160)
  Fireboats: Then and Now: Special Report (USFA-TR-146)
  Firefighter Arson: Special Report (USFA-TR-141)
  Firefighter Fatalities in the United States series [for years 1986 to 2005]
  Firefighter Fatalities Retrospective Study 1990-2000 (FA-220)
~ Four Years Later—A Second Needs Assessment of the U.S. Fire Service (October 2006) (FA-303)
~ Funding Alternatives for Fire and Emergency Services (FA-141)
~ Health and Safety Issues of the Female Emergency Responder (June 1996) (FA-162)
  HMEP Guidelines for Haz Mat/WMD Response, Planning and Prevention Training
  If You Respond to Fire on Federal Property (FA-218)
  Implementation of EMS in the Fire Service (FA-167)
  Improving Firefighter Communications: Special Report (USFA-TR-099)
  Many Faces, One Purpose: A Manager's Handbook on Women in Firefighting (FA-196)
  Many Women Strong: A Handbook for Women Firefighters (FA-195)
  National Fallen Firefighters Foundation Resource Guide
  National Fire Academy Two Day Resident Program (FA-224)
  Northwest Firefighters Mortality Study (FA-105)
  Orientation Manual for First Responders on the Evacuation of People with Disabilities (FA-235)
~ Personnel Accountability System Technology Assessment (FA-198)
~ Prevention of Self-Contained Breathing Apparatus Failures: Special Report (USFA-TR-088)
  Rapid Intervention Teams and How To Avoid Needing Them: Special Report (USFA-TR-123)
  Regional Delivery Brochure (FA-238)
~ Report of the Joint Fire/Police Task Force on Civil Unrest (February 1994) (FA-142)
  Risk Management Planning for Hazardous Materials: What It Means for Fire Service Planning: Special Report
      (USFA-TR-124)
  Risk Management Practices in the Fire Service (FA-166)
  Safe Operations of Fire Tankers (FA-248)

~Available online only

Safety and Health Considerations for the Design of Fire and Emergency Medical Services Stations (FA-168)

Small Board and Care Fire Evacuations: A Guide for the Fire Safety Professional

Strategies for Marketing Your Fire Department Today and Beyond (FA-182)

Trends and Hazards in Firefighter Training: Special Report (USFA-TR-100)

U.S. Fire Administration Training Catalog (FA-273)

The USFA Web site discusses efforts related to firefighter health and safety: *http://www.usfa.dhs.gov/fireservice/subjects/health/index.shtm*. The site that lists USFA publications in this area is *http://www.usfa.dhs.gov/applications/publications/browse.cfm?sc=14* and the site that covers research efforts in health and safety is *http://www.usfa.dhs.gov/fireservice/research/safety/index.shtm*.

In addition to ordering through the online catalog, publications may be ordered by calling the Publications Center at (800) 561–3356 between 7:30 a.m. and 5 p.m. EST/EDT. To order publications by mail, write to:

Publications Center
U.S. Fire Administration
16825 S. Seton Avenue
Emmitsburg, MD 21727

Please include your name, mailing address, daytime telephone number, date required, title(s) of the publication, and the quantity you need when ordering by phone or mail. Please include the parenthetical publication number, if given, in your request.

Publications also may be ordered online at *http://www.usfa.dhs.gov/applications/publications*

www.ingramcontent.com/pod-product-compliance
Lightning Source LLC
Chambersburg PA
CBHW081247170526
45165CB00009B/3229